Andréa Bicca Noguez Martins
Leticia Winke Dias
Aline Klug Radke

Seed physiological quality

Andréa Bicca Noguez Martins
Leticia Winke Dias
Aline Klug Radke

Seed physiological quality

Respiratory activity in corn, soybean and bean seeds

Imprint

Any brand names and product names mentioned in this book are subject to trademark, brand or patent protection and are trademarks or registered trademarks of their respective holders. The use of brand names, product names, common names, trade names, product descriptions etc. even without a particular marking in this work is in no way to be construed to mean that such names may be regarded as unrestricted in respect of trademark and brand protection legislation and could thus be used by anyone.

Cover image: www.ingimage.com

This book is a translation from the original published under ISBN 978-3-330-75883-4.

Publisher:
Sciencia Scripts
is a trademark of
Dodo Books Indian Ocean Ltd. and OmniScriptum S.R.L publishing group

120 High Road, East Finchley, London, N2 9ED, United Kingdom
Str. Armeneasca 28/1, office 1, Chisinau MD-2012, Republic of Moldova, Europe
Printed at: see last page
ISBN: 978-620-8-30292-4

SUMMARY

SUMMARY

The physiological quality of seeds is normally assessed by standard germination and vigor tests. However, in batches with a high degree of heterogeneity, these tests can present low sensitivity, making it necessary to obtain reliable results in a short period of time, ensuring speedy decision-making in relation to the physiological quality of seeds. With this in mind, the aim of this study was to separate batches of soybean, bean and corn seeds in terms of vigor, using respiratory activity. To this end, three batches of seeds from these crops were subjected to standard viability and vigor tests, such as germination, first germination count, germination speed index, electrical conductivity, seedling emergence and emergence speed index in the greenhouse, length and dry mass of the aerial part and roots of the seedlings from the germination and seedling emergence tests and respiratory activity. According to the results of the standard physiological quality tests, batch two was the most vigorous for soybean seeds and batch three for bean and maize seeds, corroborating the responses found through the seeds' respiratory activity. Therefore, it was concluded that respiratory activity makes it possible to separate batches of soybean, bean and maize seeds in terms of vigor.

Keywords: respiration, physiological quality, *Glycine max* L., *Phaseolus vulgaris* L., *Zea mays* L.

CHAPTER I

1- Seed quality analysis: an evolving activity

Seed analysis uses technical procedures to assess the quality of a batch of seeds, where quality is understood to be the set of genetic, physical, physiological and health attributes that affect a seed's ability to produce more productive plants (TILLMANN; MIRANDA, 2006). High seed quality has a direct impact on the resulting crop, in terms of population uniformity, the absence of seed-borne diseases, high plant vigor and higher productivity (POPINIGIS, 1985; CARVALHO; NAKAGAWA, 2000).

Physiological quality is normally assessed by the germination test, the aim of which is to determine the viability of a batch of seeds, the value of which can be used to compare the quality of different batches and estimate the sowing value in the field (ISTA, 1995), which takes into account the ideal conditions required by the species, showing its maximum germination capacity. However, this parameter can vary greatly under field conditions, which is why it is important to use vigor tests to obtain more consistent results for sowing (LOPES; ALEXANDRE, 2010).

By definition, a batch of seeds of the same species is formed from a defined, identified and homogeneous quantity of seeds with similar physical and physiological attributes, which is very important because it allows sampling to truly represent the entire batch of seeds (PESKE et al., 2010). Thus, in batches with a high degree of homogeneity, physiological quality can be reasonably well assessed using the standard germination test. However, for lots with a high degree of heterogeneity, the standard germination test has low sensitivity and, in this case, vigor tests better represent the performance of the lots at field level (PESKE et al., 2006), where the seeds are subjected to variations in soil moisture, radiation, competition and unfavorable conditions to express their full germination potential (HILHORST et al., 2001). In view of this, the validity of the germination test for predicting seed responses in the field, where ideal environmental conditions are unlikely to occur, has been questioned (FRIGERI, 2007). Vigor comprises a set of properties that determine the ability to emerge and the rapid development of normal seedlings, under a wide range of environmental conditions (BAALBAKI et al., 2009). Thus, its basic objective is to properly identify which batches have the greatest potential for establishing themselves in the field (MARCOS FILHO, 2005).

Identifying the type of test that offers a safe margin for seed responses in the field has been an incessant and necessary search, given that adverse conditions impose a lack of uniformity between the standard germination test and the field results, establishing the need to identify tests that provide equivalent germination conditions in the field and all the adversities that can affect their performance.

The impetus for research into seed technology is strong, with a predominance of projects aimed at methods for evaluating vigor and its relationship with seed performance, and identifying the root causes of this behavior, especially when it comes to the situation of the seed in the field.

Seed analysis is an important tool in quality control, especially from the end of the ripening period, when seeds reach their physiological maturity. Therefore, the selection of vigor tests must meet specific objectives, and it is essential to identify the characteristics evaluated by the method and their relationship with the behavior of the seeds in specific situations, such as performance after drying, storage potential, response to mechanical injuries and climatic conditions (BAALBAKI et al., 2009).

Seed lots with similar germination percentages can exhibit different responses in the field and after storage (FRIGERI, 2007). In view of this, the effort to determine the answers to the main questions regarding seed performance has become exclusively attributed to "high" or "low vigor". As a result, vigor became the main reason for the success or failure of stand establishment in the field (KRZYZANOWSKI et al., 1999). Although various studies have sought to standardize vigour tests (MARCOS FILHO et al., 2009), certain difficulties have been encountered, since this characteristic can be reflected by various variables such as germination speed, uniformity of emergence, resistance to cold, temperature, high humidity, toxic substances, among others.

In view of this, it is important to carry out tests that respond to this set of characteristics, which are related to the membrane system, enzymatic activity and the reduction of energy mechanisms (LAMARCA, 2009), and the fastest tests studied are related to these events (FREITAS, 2009) such as the degradation of cell membranes, reduced respiratory activity and decreased biosynthesis (MARCOS FILHO, 2005).

A viable and quick alternative for assessing vigor would be to subject the seeds to measurement of their respiratory activity in laboratory conditions, since respiration represents the oxidation of organic substances in a cellular system with the gradual release of energy, a metabolic process that accompanies the rehydration of the seed, which rises from minimal values to very high levels a few hours after the start of soaking (FERREIRA; BORGHETTI, 2004; MARENCO; LOPES, 2007).

Among the different ways of checking the physiological quality of seeds, the respiration process has received special attention due to the high relationship between this phenomenon and seed quality (MENDES et al., 2009; AUMONDE et al., 2012; MARINI et al., 2012). With this in mind, the aim was to evaluate respiratory activity in order to separate batches of soybean, bean and corn seeds in terms of vigor.

2- Seed quality and physiological potential

The history of agriculture shows that the first contacts between man and the physiology of seeds were established from the moment the possibility of using them to propagate plants was discovered in the 19th century BC (ABRATES, 1999). In this situation, as well as causing profound positive changes in human life habits, the beginning of the use of seeds for the establishment of crops, with a view to food production, also became a source of concern (PERES, 2010).

Thus, with the domestication of plant species, new challenges have arisen with the need to determine the most favorable times for sowing (MARCOS FILHO, 2005). Seed is an indispensable input due to its role in the agricultural chain and in human history, both socially and economically, contributing to quantitative and qualitative increases in productivity. Therefore, the use of high quality seeds is a key factor in the success of any crop (GASPAR and NAKAGAWA, 2002).

The procedures adopted in seed quality control programs are based both on previous knowledge of recommendations from research or practical experience, and on the collection of data that allows problems to be detected and appropriate solutions to be proposed (KRZYZANOWSKI et al., 1999).

Physiological potential provides information on the viability and vigor of a batch of seeds, the term potential being translated as virtuality or a set of abilities to perform tasks and produce results (MARCOS FILHO, 2005). The test commonly used to determine the viability of seeds is germination, the main objective of which is to obtain information on the importance of seeds for sowing and to compare the quality of different batches (LIMA et al., 2006).

It is conducted under optimum conditions in order to provide maximum germination of the sample analyzed. These conditions refer to the availability of water, aeration and temperature (MARCOS FILHO et al., 1987; BRASIL, 2009). The germination test has at least two aspects in mind: to provide information on the potential of a batch to germinate under favourable environmental conditions and to present a high degree of standardization, with ample possibility of repeating the results, within reasonable tolerance levels, as long as the established instructions are followed (MARCOS FILHO, 1999; BRASIL, 2009).

However, the germination test can overestimate the physiological potential of seeds because it does not evaluate the physiological, biochemical, physical and cytological changes related to the deterioration process, and does not allow seed lots to be differentiated in the field and in storage in terms of vigor (ABRANTES et al., 2010). For this reason, research has been carried out to develop methods for assessing seed vigor (KIKUTI et al., 1999; AVILA et al., 2007; OHLSON et al., 2010).

The vigor of a seed comprises a set of properties that determine the ability to emerge and the rapid development of normal seedlings under a wide range of environmental conditions (BAALBAKI et al., 2009). Thus, its basic objective is to properly identify which batches have the greatest potential to survive and generate good productivity under field conditions (MARCOS FILHO, 2005).

In the United States and Canada, from 1976 to 1990, there was a significant increase in the use of vigor tests in seed analysis laboratories, at which time the Electrical Conductivity test was not mentioned as a vigor test (TEKRONY, 1983; FERGUSON, 1993). This test, together with the accelerated ageing and cold tests, were the subject of study by the vigor committee of AOSA (*Association of Official Seed Analysts, Inc.*), which, in the period 1983-1991, were considered the three most promising vigor tests (McDONALD, 1993).

Researching the use of vigor tests among ISTA (*International Seed Testing Association*) member laboratories, Hampton et al. in 1992 considered the electrical conductivity test as a recommended test for evaluating pea seed vigor in Europe and New Zealand. As a result, electrical conductivity and the accelerated ageing test became the only two vigor tests recommended by the ISTA vigor committee (HAMPTON and TEKRONY, 1995).

In Brazil, Krzyzanowski et al. (1991), evaluating the use of vigor tests as a routine among seed analysis laboratories, concluded that, despite being fundamental, these tests have a lot to evolve in order to effectively participate in the quality control programs of seed industries. In the particular case of the electrical conductivity test, it can be said that its use is still very restricted to certain situations, especially those directly related to research (PERES, 2010).

According to ABRATES (Associaçâo Brasileira de Tecnologia de Sementes, 1999), there are three criteria for seed quality: germination, purity and health. These can be determined by daily analysis in seed laboratories and are of great importance for assessing quality on the market, but they are not considered to be the most efficient.

Seed analysis is an important tool in quality control, especially from the end of the ripening period, when the seeds reach physiological maturity. Thus, the opening up of new agricultural frontiers and the increase in seed production in Brazil in recent years have led seed companies to seek technical improvements in their activities, which basically aim to increase productivity associated with an increase in the quality of the harvested product.Both ISTA and AOSA adopt the procedure of evaluating the most appropriate methodology for inclusion in the Rules for Seed Analysis, through benchmarking tests carried out in different laboratories, under the coordination of specific committees where purity, germination, moisture content and other characteristics are assessed (ABRATES, 1999).

To do this, a certain number of samples are sent to the laboratories participating in the program, accompanied by instructions that must be followed by the analysts. Once the Committee has the results, it interprets them, verifying compatibility between the laboratories, detecting problems, diagnosing the situation and scheduling new stages of testing, until the level of standardization is satisfactory and allows the methodology to be recommended. There is a careful process of standardization and quality control which aims to match and determine the best method for vital seed monitoring (ABRATES, 1999).

The main purpose of seed analysis is to determine the quality of a seed lot and, consequently, its value for sowing (EMBRAPA, 2008). Analysis is characterized by the detailed and critical examination of a sample, with the aim of assessing its quality, in order to be used in research work, to identify quality problems and their causes.

In the United States, most major crop seed companies have used vigor tests to identify lots that don't meet internal quality standards, such as classifying (ranking) lots into different levels of physiological quality, assessing the potential for forming regulatory stocks (*carry over*), making decisions about marketing, seeking to market first those batches that meet germination standards but have lower vigor, as well as providing consumers with information about the physiological quality of the batches (FRIGERI, 2007).

There is an international consensus among researchers, technologists and seed producers about the importance of seed vigor and the need to evaluate it. Information on vigor is even more important for seeds of greater commercial value, such as vegetables, which may have been pelleted (covered in films) and physiologically preconditioned, as is done in other countries. In addition, because they have less stored reserves, they are more prone to a reduction in vigor after physiological maturity (KIKUTI and MARCOS FILHO, 2012). These species are grown intensively and must be established using seeds that germinate quickly and uniformly and, therefore, have high physiological quality.

For this reason, the search for methods that are satisfactory in predicting the physiological quality of seeds by means of vegetable seed vigor serves as a methodological update for other crops, such as hybrid corn seeds, generating technological progress in the search for quality in the determination of vigor tests (PERES, 2010).

In this context, vigor tests are useful in seed production programs for evaluating the physiological potential of lots with similar germination, making it possible to differentiate lots based on the potential for seedling emergence in the field; evaluating storage potential; degree of deterioration; post-ripening quality control; physiological quality, serving as a tool to aid selection methods during plant breeding; as well as making it possible to assess the effects of mechanical and thermal injuries, treatment with fungicides and other adverse pre- and post-harvest factors (MARCOS FILHO, 1999).

Seed technology has sought to improve germination and vigor tests in order to obtain results that express the effective performance of seeds in the field. In this case, interest in vigor tests has been growing, especially in the internal quality control programs of seed companies (KIKUTI and MARCOS FILHO, 2012).

Often batches of seeds with similar germination percentages can show different responses in the field and/or in storage (FRIGERI, 2007). This loss of germination potential is an important indication of loss of quality, but it is the last consequence, i.e. the final event in this process. For this reason, the use of vigor tests is very important in monitoring seed quality from maturity onwards, as the drop in vigor precedes the loss of viability (DIAS and MARCOS FILHO, 1995).

3- Vigor tests

Despite their technological differences, vigor tests aim to detect significant differences in the physiological potential of seed lots with similar germination (LIMA et al., 2006; DUTRA and MEDEIROS FILHO, 2008), classifying them into levels of vigor, especially in proportion to the response of seedling emergence in the field (MARCOS FILHO, 1999). Therefore, the selection of vigor tests must meet specific objectives, making it important to identify the characteristics evaluated by the test and their relationship with seed responses in specific situations, such as performance after drying, storage potential, response to mechanical injuries and climatic conditions.

According to McDonald (1993), vigor tests can be classified as physical, physiological, biochemical and stress resistance. Physical tests evaluate morphological or physical characteristics of the seeds that may be associated with vigor, such as size, density, seed color and X-ray testing. Tests characterized as physiological are based on specific physiological activities that have their manifestation dependent on vigour, such as the first germination count, the germination speed index or seedling emergence. Biochemical tests assess changes in metabolism related to seed vigor, including tetrazolium and electrical conductivity tests. Lastly, stress resistance tests analyze the behavior of seeds when exposed to unfavorable environmental conditions, especially accelerated aging, controlled deterioration, cold, low temperature germination and water submersion (MARCOS FILHO, 2005).

The main challenge in vigor testing research is to identify suitable parameters that are common to seed deterioration, so that the easier it is to identify the loss of germination capacity, the more promising the test will be, thus providing complementary information to that obtained through the standard germination test (AOSA, 1983).

Seed vigor, as defined by the *International Seed Testing Association* (ISTA, 1995), is an index of the degree of physiological deterioration and/or mechanical integrity of a batch of highly germinated seeds, representing a broad ability to establish in the environment. This definition of vigor is similar to that formulated by the *Association of Official Seed Analysts* (AOSA, 1983).

Vigor tests help to detect this information and are therefore useful in making decisions about the fate of a batch of seeds. Among the vigor tests available, it is worth highlighting the electrical conductivity test, which is a quick and objective vigor test that can be easily carried out by various seed analysis laboratories, with minimal expenditure on equipment and staff training (MARCOS FILHO et al., 2009).

The results of vigor tests are comparative, and it is not possible to quantify seed vigor in the same way that soil fertility is quantified, because they are all non-measurable characteristics. In fact, the result of 60% normal seedlings in the accelerated ageing, cold and first germination count tests, among others, means nothing if it is not compared with that obtained for another sample of the same species and cultivar. Thus, expressions such as 70% vigor are incorrect and should not be used (ISTA, 2001).

The impossibility of quantifying vigor creates difficulties both in understanding its meaning and in comparing information obtained from different tests. Research has therefore sought to translate and establish indices that help to facilitate the interpretation and use of the results (PERES, 2010).

Evaluating the physiological potential of seeds is essential as a basis for the production, distribution and marketing of seed lots. Producing companies and seed analysis laboratories must therefore use tests that offer reproducible, reliable results and that reliably indicate the quality of a batch of seeds, especially with regard to vigor (FRIGERI, 2007). To this end, several studies have been carried out with the aim of finding or improving tests that meet these characteristics. Delouche (1965) based himself on information obtained by Crocker and Graves (1915), according to which the death of seeds during storage was caused by the coagulation of proteins and that heating accelerated this process; both researchers also suggested that germination tests, conducted after relatively rapid exposure of seeds to high temperatures (50-100°C), could be useful for obtaining more rapid information on longevity.

A batch of seeds of the same species is formed from a defined, identified and homogeneous quantity of seeds with similar physical and physiological attributes. However, batches, even from the same production area, may show imperfect homogeneity in terms of germination and vigor, which is related to edaphic variations in topography and fertility, or may even be influenced by harvesting interspersed with rainy days (PESKE et al., 2006).

Vigor analyses allow lots to be ranked, enabling them to be marketed according to local growing conditions, so that lots with greater vigor can be destined for regions with greater environmental limitations during the sowing period (PESKE et al., 2006).

Various vigor tests are available and differ in terms of methodology, time and ease of execution. The most studied are those related to the initial events in the deterioration sequence (MARCOS FILHO, 1999), such as the degradation of cell membranes, reduction of respiratory activity and the separation of seed lots in terms of vigor (ABRANTES et al., 2010).

The first germination count test is based on the principle that the samples with the highest percentages of normal seedlings in the first count, established by the Rules for Seed Analysis (BRASIL, 2009), for each crop will be the most vigorous, which correlates with the germination speed index, but may have a better response than the latter, as reported by Brown and Mayer (1986), thus reinforcing the claim that this test is of great interest for assessing seed vigor, taking into account its practicality and execution time.

As the germination test itself is used to carry out both of the tests mentioned above, it is sufficient to follow the rules of the Rules for Seed Analysis (BRASIL, 2009) to carry them out, where the uniformity and speed of seedling emergence are the most important components within the current conceptualization of seed vigor, with the evaluation of seedling growth being considered a logical and specific test for testing vigor, as well as the evaluation of the length of normal seedlings (AOSA, 1983).

The assessment of dry mass and seedling length is related to the speed of germination, taking into account that batches with more vigorous seeds will produce seedlings with higher rates of development and biomass gain due to their greater capacity for adaptation, using the reserves of storage tissues as a supply for tissue differentiation, promoting the growth and development of the embryonic axis and, consequently, of the seedling (DAN et al., 1987).

Methods for assessing vigor can be classified as direct when carried out in the field or even under laboratory conditions that simulate adverse field factors; or indirect when carried out in the laboratory, assessing the physical, physiological and biochemical characteristics that express the quality of the seeds (FERREIRA and BORGHETTI, 2004). In general, low seed vigor is associated with reductions in the speed and unevenness of emergence, as well as reductions in the initial size of seedlings, dry mass accumulation, leaf area and, consequently, crop growth rates (SCHUCH et al., 2000; MACHADO, 2002; HOFS, 2004; KOLCHINSKI et al., 2005). The cause of failure or reduction in the speed of emergence is often attributed to low vigor, associated with the process of seed deterioration (ROSSETTO et al., 1997).

As reported, various vigor tests are available and differ in terms of methodology, time and ease of execution. The most studied are those related to the initial events of the deterioration sequence (DELOUCHE; BASKIN, 1973), such as the degradation of cell membranes and a reduction in respiratory activity, allowing seed lots to be separated in terms of vigor (ABRANTES et al., 2010).

Seed vigor and deterioration are physiologically linked, being reciprocal aspects of quality, where deterioration has a negative connotation, while vigor has a positive connotation, as they are inversely proportional (DELOUCHE, 2002). The seed does not begin the deterioration process until it reaches physiological maturity, since before this period it is not yet an independent unit of the mother plant. However, unfavorable environmental conditions during maturation can lead to the formation of seeds with poor physiological potential (MARCOS FILHO, 2005).

Among the tests that evaluate the initial events of deterioration, the electrical conductivity test was proposed by Matthews and Bradnock (1967) to estimate the vigor of pea seeds. This test assesses the amount of electrolytes released by the seeds during the soaking period, which is directly related to the integrity of the cell membranes (MATTHEWS; POWELL, 1981). Poorly structured membranes and damaged cells are generally associated with the process of seed deterioration and, therefore, with seeds of low vigor (AOSA, 1983). Seed vigor is therefore considered to be inversely proportional to electrical conductivity (VIEIRA, 1994; VIEIRA and KRZYZANOWSKI, 1999).

Respiration is the first metabolic activity that accompanies the rehydration of the seed, and the increase in respiration varies from minimal to very high levels shortly after the start of soaking (POPINIGIS, 1977; FERREIRA; BORGHETTI, 2004). This process involves the oxidation of organic substances in a cellular system with the gradual release of energy through a series of reactions, with molecular oxygen as the final electron acceptor. Respiratory substrates can be carbohydrates such as starch, sucrose, fructose, glucose and other sugars or even lipids, especially triglycerides, organic acids and proteins (MARENCO; LOPES 2007; TAIZ; ZEIGER, 2009). For this reason, among the various procedures used to determine seed vigor, one of the alternatives would be to subject them to measurement of respiratory activity in laboratory conditions (MENDES et al., 2009; AUMONDE et al. 2012; MARINI et al., 2012).

The increase in the seed's respiratory activity can be assessed by the amount of carbon dioxide (CO_2) eliminated, the amount of oxygen (O_2) absorbed or the respiratory quotient (RQ). The seed's respiration rate is influenced by its moisture content, temperature, membrane permeability, oxygen tension and light (POPINIGIS, 1977). Respiration implies the loss of dry mass and gas exchange, and the methods used to measure respiration are based on determining these characteristics. However, measuring the change in dry mass requires a large amount of material, as well as being considered a somewhat time-consuming analysis to obtain the result, given that the plant material must be completely dried in an oven (MARENCO; LOPES, 2007).

Methods based on gas exchange are more sensitive, require fewer materials and are non-destructive. They can consist of manometric measurement of O_2 consumed, for example, the Warburg respirometer and the Clark electrode (potentiometry), the measurement of CO_2 released, using physical methods such as the infrared gas analyzer (IRGA), or physical-chemical methods based on the retention of CO_2 in a base and quantification by titrimetry, calorimetry or conductivimetry (MAESTRI et al., 1998).

Among the different ways of checking the physiological quality of seeds, the respiration process has received special attention due to the high relationship between this phenomenon and seed quality (MENDES et al., 2009, AUMONDE et al., 2012). Bearing in mind that the greatest interest today when assessing the physiological quality of seeds is to obtain reliable results in a relatively short period of time, it is hoped that the agility of this assessment will allow decisions to be made quickly during different stages of seed production, especially between the ripening stage and future sowing (DIAS and MARCOS FILHO,

1996).

Tests for the rapid assessment of viability or vigor are important components in seed quality control programs and make it possible to obtain information quickly, by discarding lots of inferior quality during the reception phase at the processing plant and rationalizing handling (MARCOS FILHO, 2005).

In this context, and bearing in mind that the analysis of the physiological quality of seeds should be seen as a dynamic activity that constantly evolves, both by improving the means available for its evaluation and by incorporating new methods (NOVEMBRE, 2001), it is extremely important to prove the efficiency of new analyses, such as the breath test, which aims to separate batches of seeds in terms of vigor.

REFERENCES

ABRANTES, F.L.; KULCZYNSKI, S.M.; SORATTO, R.P.; BARBOSA, M.M.M. Nitrogen in top dressing and physiological and health quality of *millet* seeds (*Panicum miliaceum* L.). **Revista Brasileira de Sementes**, v.32, n.3, p.106-115, 2010.

AMARAL, A.S. & PESKE, S.T. pH of the exudate to estimate, in 30 minutes, the viability of soybean seeds. **Revista Brasileira de Sementes**, v.6, n.3, p.85-92, 1984.

AOSA. Association of Official Seed Analysts. **Seed vigor testing handbook.** East Lansing, AOSA, 1983. 88p.

AUMONDE, T.Z.; MARINI, P.; MORAES, D.M. de; MAIA, M.S.; PEDÓ, T.; TILLMANN, M.A.A.; VILLELA, F.A. Classification of the vigor of kidney bean seeds by respiratory activity. **Interciência**, v.37, n.1,   p.55-58, 2012.

AVILA, M.R.; BRACCINI, A.L.; SCAPIM, C.A.; MANDARINO, J.M.G.; ALBRECHT, L.P.; VIDIGAL FILHO, P.S. Yield components, content of isoflavones, proteins, oil and quality of soybean seeds. **Revista Brasileira de Sementes**, v. 29, n.3, p.111-127, 2007.

BAALBAKI, R.; ELIAS, S.; MARCOS FILHO, J., McDONALD, M.B. Seed vigor testing handbook. **Association of Official Seed Analysts** (Contribution, 32 to the Handbook on Seed Testing), 346p., 2009.

BRAZIL. Ministry of Agriculture, Livestock and Supply. **Rules for Seed Analysis**. Secretariat for Agricultural Defense. Brasilia: MAPA/ACS, 2009. 398p.

BROWN, R.F.; MAYER, D.G. A critical analysis of Maguire's germination rate index. **Seed Science and Technology**, v.10, p.101-110, 1986.

CARVALHO, N.M.; NAKAGAWA, J. (Ed.) **Seeds: science, technology and production**. 4 ed. Jaboticabal: FUNEP, 2000. 588p.

DAN, E.L.; MELLO, V.D.C.; WETZEL, C.T.; POPINIGIS, F.; ZONTA, E.P. Dry matter transfer as a method for evaluating soybean seed vigor. **Revista Brasileira de Sementes**, v.9, n.2, p.45-55, 1987.

DELOUCHE, J.C. Seed Germination, Deterioration and Vigor. Cover story Nov/Dec. **Seed News magazine**, v.6, n.6, 2002.

DELOUCHE , J.C.; BASKIN, C.C. Accelerated aging techniques for predicting the relative storability of seed lots. **Seed Science and Technology**. v.1, p.427-452,1973.

DELOUCHE, J.C. An accelerated aging technique for predicting the relative storability of crimson clover and tall fescue seed lots. **Agronomy Abstracts**, v.57, p.40,1965.

DIAS, D.C.F.S.; MARCOS FILHO, J. Electrical conductivity tests for evaluating the vigor of soybean seeds (*Glycine max* (L.) Merrill). **Scientia Agricola**. v.53, p.31-42, 1996.

DIAS, D.C.F.S.; MARCOS FILHO, J. Vigor tests based on cell membrane permeability: II Potassium leaching. **Informativo ABRATES**, Brasilia, v.5, n.1, p.37-41, 1995.

DUTRA, A.S.; MEDEIROS FILHO, S. Controlled deterioration test in the determination of vigor in cotton seeds. **Revista Brasileira de Sementes**, v.30, n.1, p.19-23, 2008.

BRAZILIAN AGRICULTURAL RESEARCH COMPANY. National Rice and Bean Research Center. **Technical recommendations for growing beans**. 2.ed. Goiânia: Embrapa CNPAF, 40p. (Technical Circular, 13), 2008.

FERGuSON, J.M. AOSA Perspective of seed vigor testing. **Journal Seed Science and Technology**, v.17, n.2, p.101-4. 1993.

FERREIRA, A.G.; BORGHETTI, F. **Germination**: from basic to applied. Porto Alegre: Artmed, 2004. 323p.

FREITAS, R. A. Deterioration and storage of vegetable seeds. In: NASCIMENTO, W. M. (Ed.). Vegetable seed technology. Brasilia, DF: Embrapa hortaliças, 2009. p. 155-184.

FRIGERI, T. **Interference of pathogens in the results of vigor tests on bean seeds**. 77f. Dissertation (Master's Degree in Agronomy). Faculty of Agricultural and Veterinary Sciences. Universidade Estadual Paulista, Jaboticabal, 2007.

GASPAR, C.M.; NAKAGAWA, J. Electrical conductivity test as a function of soaking period and temperature for millet seeds. **Revista Brasileira de Sementes**, v.24, n.2, p.82-89, 2002.

HAMPTON, J.G.; TEKRONY, D.M. Handbook of vigour test methods. 3ed.

Zurich:ISTA, 117p , 1995.

HAMPTON, J.G.; JOHNSTONE, K.A.; EUA-UMPON, V. Bulk conductivity test variables for mungbean, soybean and French bean seed lots. **Seed Science and Technology**, v.20, n.3, p.677-686, 1992.

HILHORST, H.W.M.; BEWLEY, J.D.; CASTRO, R.D.; SILVA, E.A.A. **Advanced course in seed physiology and technology**. Lavras: UFLA, 2001. 74p.

HOFS, A.; SCHUCH, L.O.B.; PESKE, S.T. Emergence and growth of rice seedlings in response to seed physiological quality. **Revista Brasileira de Sementes**. v. 26, n.1, p.92-97, 2004.

Informativo ABRATES, Londrina, v.4, n.3, 1999. 218p.

ISTA. Report of the ISTA committees: The 26th ISTA Congress, 2001. **Seed Science and Technology**, v.29, 2001.

ISTA.INTERNATIONAL SEED TESTING ASSOCIATION. **Handbook of vigour test methods**. 3.ed. Zürich: ISTA, 1995. 116p.

KIKUTI, A.L.P.; VON PINHO, E.V.R.; REZENDE, M.L. Studies of methodologies for conducting the cold test on maize seeds. **Revista Brasileira de Sementes**, v.21, n.2, p.175-179, 1999.

KIKUTI, A.L.P.; MARCOS FILHO J. Vigor tests on lettuce seeds. **Horticultura Brasileira**, v.30, p.44-50, 2012.

KOLCHINSKI, E.M.; SCHUCH, L.O.B; PESKE, S.T. Seed vigor and intraspecific competition in soybean. **Ciência rural**, v.35, n.6, p.1248-1256, 2005.

KRZYZANOWKI, F.C.; VIEIRA, R.D. Controlled deterioration. In: KRZYZANOWKI, F.C.; VIEIRA, R.D.; FRANÇA NETO, J.B. (Ed.). **Seed vigor**: concepts and tests. Londrina: ABRATES, p.61-68, 1999.

KRZYZANOWSKI, F.C.; FRANÇA NETO, I.B.; HENNING, A.A. **Report on vigor tests available for major crops**. Inf. ABRATES, Londrina, v.1, n.2, p.15-50. 1991.

LAMARCA, E.V. **Respiration rates and deterioration speed of *Caesalpinia echinata* Lam. seeds as a function of hydric and thermal variations.** Dissertation (Master's Degree in Plant Biodiversity and Environment), Instituto de Botânica de Sâo Paulo, Sâo Paulo. 98p. 2009.

LIMA, T.C.; MEDINA, P.F.; FANAN, S. Evaluation of wheat seed vigor by the accelerated aging test. **Revista Brasileira de Sementes**, v.28, n.1, p.106-113, 2006.

LOPES, J.C.; ALEXANDRE, R.S. Germination of forest species seeds. In: José Franklim Chichorro; Giovanni de Oliveira Garcia; Maristela de Oliveira Bauer; Marcos Vinicius Winckler Caldeira. (Org.). **Tópicos em Ciências Florestais**. 1 ed., v.1, p.21-56, 2010.

MAESTRI, M.; ALVIM, P. de T.; SILVA, M.A.P. **Fisiologia vegetal**; exercicios prâticos. Viçosa: UFV, 1998. 91p. (Cadernos didâticos, 20).

MCDONALD, M.B. A review and evaluation of seed vigor tests. **Proceedings of the Associations of Official Seed Analysts**, v.65, p.109-139, 1993.

MACHADO, R.F. **Performance of black oats (*Avena sativa* L.) as a function of seed vigor and plant population.** 46 f, Dissertation (Postgraduate course in Seed Science and Technology) "Eliseu Maciel" School of Agronomy, **Federal** University of Pelotas, **2002.** Federal University of Pelotas, 2002.

MARCOS FILHO, J.; KIKUTI, A. L. P.; LIMA, L. B.Methods for evaluating the vigor of soybean seeds, including computerized image analysis. Revista Brasileira de Sementes, v. 31, n. 1, p. 102-112, 2009.

MARCOS FILHO, J. **Fisiologia de sementes de plantas cultivadas**. Piracicaba: FEALQ, v.1, 2005. 495p.

MARCOS FILHO, J. Vigor tests: Importance and use. In: KRZYZANOWSKI, F.C., VIEIRA, R.D., FRANCA NETO, J.B. (Ed.). Seed vigor: Concepts and Tests. **Londrina: ABRATES**, ch.1, p.1-21. ,1999 .

MARCOS FILHO, J.; CICERO, S.M.; SILVA, W.R. da. **Avaliaçâo da qualidade de sementes**, Piracicaba: FEALQ, 1987. 230p.

MARCOS FILHO, J. **Fisiologia de sementes de plantas cultivadas**. Piracicaba: FEALQ, v.1, 2005. 495p.

MARENCO, R.A.; LOPES, N.F. **Plant physiology**: photosynthesis, respiration, water relations and mineral nutrition. 2.ed. Viçosa: UFV, 2007. 469p.

MARINI, P.; MORAES, C.L.; MARINI, N.; MORAES, D.M. de; AMARANTE, L. Physiological and biochemical changes in rice seeds subjected to heat stress. **Revista Ciência Agronômica**, v.43, n.4, p.722-730, 2012.

MATTHEWS, S.; BRADNOCK, W.T. The detection of seed samples of wrinkled-seeded peas (*Pisum sativum* L.) of potentially low planting value. **Proceedings of International Seed Testing Association**, v. 32, p.553-563, 1967.

MATTHEWS, S.; POWELL, A.A. Electrical conductivity test. In: PERRY, D.A. (Ed.). **Handbook of vigour test methods.** ISTA, p.37-41, 1981.

MENDES, C.R.; MORAES, D.M.; LIMA, M.G.S.; LOPES, N.F. Respiratory activity for the differentiation of vigor on soybean seed lots. **Revista Brasileira de Sementes**, v.31, p.171-176, 2009.

NOVEMBRE, A.D.L.C. **Avaliaçâo da qualidade de sementes**, Londrina. 2001.

OHLSON, O.C.; KRZYZANOWSKI, F.C.; CAIEIRO, J.T.; PANOBIANCO, M. Accelerated aging test on wheat seeds. **Revista Brasileira de Sementes**, v.32, n.4, p.118-124, 2010.

PERES, W.L.R. **Vigor tests on corn seeds.** 50 f.

Dissertation (Master's Degree in Agronomy). Faculty of Agricultural and Veterinary Sciences - UNESP, 2010.

PESKE, S.T.; BARROS, A.C.S.A.; SCHUCH, L.O.B. Benefits and obtaining high quality seeds. Seed news, year 14, n.5, 2010.

PESKE, S.T.; LUCCA FILHO, O.A.; BARROS, A.C.S.A. **Sementes:** fundamentos Cientificos e tecnológicos. 2.ed. Pelotas: Ed. Universitâria/UFPel, 470p., 2006.

POPINIGIS, F. **Seed physiology**. Brasilia, Agiplan, 1977. 289p.

POPINIGIS, F. **Seed physiology**. Brasilia, AGIPLAN, 1985. 289p.

ROSSETTO, C.A.V.; MARCOS FILHO, J. Comparison between accelerated aging and controlled deterioration methods for evaluating the physiological quality of soybean seeds. **Scientia Agricola**, Piracicaba, v.52, n.1, p.123-131, 1995.

ROSSETO, C.Q.V.; NOVEMBRE, A.D.C.; MARCOS FILHO, J.; SILVA, W.R.; NAKAGAWA, J. Effect of substrate water availability on the physiological quality and initial water content of soybean seeds during germination. **Scientia Agricola**, Piracicaba, v.54, n.1/2, p.97-105, 1997.

SCHUCH, L.O.B.; NEDEL, J.L.; ASSIS, F.N.; MAIA, M.S. Field emergence and initial growth of black oat in response to seed vigor. **Revista Brasileira de Agrociência,** v.6, n.2, p.97-101, 2000.

TAIZ, L.; ZEIGER, E. **Plant Physiology**. 4th ed. Artmed, Porto Alegre, 820p. 2009.

TEKRONY, D.M. Seed vigor testing. **Journal of Seed Technology**. v.8, n.1, p.55-60, 1983.

TILLMANN, M.A.; MIRANDA, D. In: PESKE, S.T.; LUCCA FILHO, O.; BARROS, A.C.A. **Sementes: Scientific and technological foundations**. Pelotas, 470p. 2006.

VIEIRA, R.D. Electrical conductivity test. In: VIEIRA, R.D.; CARVALHO, N.M. **Testes de vigor em sementes**. Jaboticabal: Funep, p.103-139, 1994.

VIEIRA, R.D.; KRZYZANOWSKI, F.C. Electrical conductivity test. In: KRZYZANOWSKI, F.C.; VIEIRA, R.D.; FRANÇA NETO, J.B. (eds). Seed vigor: concepts and tests. Londrina: **ABRATES**. CAP.4, p.1-26, 1999.

CHAPTER II

1   - Respiratory activity for separating batches of seeds

Brazil is currently one of the world's leading producers and exporters of various agricultural products such as soybeans (*Glycine max* L.), beans (*Phaseolus vulgaris* L.) and corn (*Zea mays* L.). For the 2012/2013 harvest, it is estimated that the production of these grains should be the highest in history, reaching 185 million tons (Conab, 2013), so it is clear that Brazilian seed production occupies a prominent place in the world, with quality control being of fundamental importance within the scenario of technological evolution, which is constantly driven by market competitiveness (AVILA et al., 2007).

For this reason, the perception of the value of seed grows with each harvest, making it necessary to improve the techniques and methods of seed analysis in order to increase quality and, consequently, productivity, which is of concern to all the segments that make up the agricultural production chains (ABRANTES et al., 2010).

Within this context, seed analysis assesses the quality of a batch of seeds using technical procedures. Quality is understood to be the set of genetic, physical, physiological and health attributes that influence the ability to produce plants with greater productivity (MARCOS FILHO et al., 2006). A batch of seeds of the same species is formed from a defined, identified and homogeneous quantity of seeds with similar physical and physiological attributes. However, even batches from the same production area can be heterogeneous in terms of germination and vigor (AUMONDE et al., 2012).

Under suitable environmental conditions, the germination test is officially used to determine seed quality, however, as well as being a time-consuming method, it does not provide information on the potential performance of seeds when environmental conditions deviate from the most suitable (MARCOS

FILHO and KIKUTI, 2006; BARBIERI et al., 2012), making it necessary to use complementary tests that identify differences associated with the performance of seed lots during storage or after sowing (MARCOS FILHO et al., 2009), allowing greater differentiation between lots quickly and ensuring early decision-making during the seed production stages, which reduces risks and losses. Therefore, it is essential to use vigor tests that are accurate, easy to perform, quick, low-cost and highly related to routine analyses in seed analysis laboratories (DUTRA and VIEIRA, 2006).

Based on the above, and knowing that at the start of the germination process, with the rehydration of the seed, respiration is the first metabolic activity to be rapidly activated to high levels, a few hours after the start of soaking, accelerating metabolism and the activation of respiratory and hydrolytic enzymes (HOFS et al., 2004), the verification of the physiological quality of seeds through the respiration process has deserved special attention, due to the high relationship observed between this phenomenon and the quality of the seeds of some crops, such as millet (AUMONDE et al., 2012) and sunflower (DODE et al., 2012).

However, little information on the relationship between seed respiration activity and seed quality is available in the literature, which is not enough to characterize the respiration test as a standard for separating seed lots in terms of vigor. With this in mind, the aim was to separate seed lots of different crops (soybeans, beans and corn) in terms of vigor by means of respiratory activity.

2   - Determination of respiratory activity

The work was carried out at the Seed Physiology Laboratory and in a greenhouse at the Botany Department of the Federal University of Pelotas (Capâo do Leâo, RS, Brazil). We used three batches of soybean seeds, cultivar NA 4990RG, obtained from a private company, three batches of bean seeds, cultivar Guapo Brilhante, provided by the Temperate Climate Agricultural Research Center - CPACT-EMBRAPA and three batches of corn seeds, also obtained from a private company.

In order to characterize the physiological quality of the seed batches and determine their respiratory activity, the seeds were subjected to the following evaluations, in accordance with the Seed Analysis Rules (BRASIL, 2009) according to the following tests: germination test (G), carried out with four replicates of 200 seeds (four subsamples of 50 seeds), totalling 800 seeds per batch of each crop. The substrate used was a roll of special germination paper (germitest®), moistened with distilled water in the proportion of 2.5 times its initial mass and kept in a germinator at 25°C. The results were expressed as a percentage of germinated seeds, with soybeans being counted at eight, beans at nine and corn at seven days after sowing; the first germination count (PCG) was carried out together with the germination test, with the first count for soybeans and beans being carried out at five days after sowing (DAS), and for corn at four DAS. The results were expressed as a percentage of normal seedlings for each batch.

The germination speed index (GVI) was carried out according to Maguire (1962), in conjunction with the germination test, where daily counts were taken from the time the root protruded through the seed coat, until the number of emerged seedlings remained constant. The last day of counting for this test was the same as for the germination test.

The electrical conductivity (EC) test was carried out with four subsamples of

25 seeds for each replication, four replications for each batch. First, the mass of the seeds was determined and they were placed in beakers with 80 mL of deionized water and kept in a germinator at 25°C. After three and 24 hours of soaking, readings were taken on a Digimed CD-21 bench conductivity meter, with the results expressed in μS cm g^{-1-1} of seeds (KRZYZANOWSKI, 1991).

For seedling emergence in the greenhouse (E), the seeds were sown in perforated plastic trays using washed sand as a substrate. Four replicates of 200 seeds were used, divided into four 50-seed subsamples for each batch, which were evaluated at 21 DAS for the number of emerged seedlings (BRASIL, 2009). The speed of emergence index in the greenhouse (IVE) was carried out in conjunction with the emergence test through daily counts starting from seed germination, until the number of emerged seedlings remained constant, and was calculated according to Maguire (1962).

At the end of the germination and seedling emergence tests, the initial growth of the seedlings was analyzed by measuring the length of the aerial part (CPA) and roots (CR) of 40 seedlings per replication, with a millimeter ruler and the results expressed in mm seedling^{-1} (KRZYZANOWSKI *et al.*, 1991). The dry mass of the aerial part (MSPA) and roots (MSR) was determined gravimetrically, after drying in an oven at 70±1°C until a constant mass was obtained, and the results were expressed in mg seedling $.^{-1}$

For the respiratory activity (RA) determined in the Pettenkofer apparatus, 100 g of seeds of each species were used, which were soaked for 60 minutes in 80 mL of distilled water to accelerate the respiratory process. The release of CO_2 from the seeds was measured according to the methodology described by Moraes *et al.* (2012). Respiratory activity was calculated using the following equation: $N \times D \times 22/g$ seed, where N represents the normality of the acid used (HCl 0.4 N), D the difference between the volume of HCl used in the titration of the blank test and the volume of HCl used in the titration of the sample and 22 the weight in grams of CO_2 for each molecule of HCl used in the titration. The results were expressed as mg CO_2 released mg^{-1} seed h $.^{-1}$

The experimental design was entirely randomized, with four replications. The data was submitted to analysis of variance and the means were compared using the Tukey test with a 5% probability of error using the WINSTAT software (MACHADO and CONCEIÇÂO, 2007).

Viability, determined by the germination percentage, showed a significant difference for the soybean and bean lots, with lot two being superior for soybean seeds and lot three for beans (Figure 1A). The same was not true for corn seeds, where the germination test showed no significant difference between the lots, ranging from 99.75% to 100% (Figure 1A).

The first germination count (Figure 1B) and the germination speed index (Figure 1C) evaluated on soybean and bean seeds corroborated the germination test, highlighting the same lots as the most vigorous (lots two and three, respectively). These same variables made it possible to differentiate between batches of corn seeds, with batch three being superior to the others (Figure 1B and 1C, respectively). Dutra et al. (2008) used these same variables to differentiate batches of cotton seeds. However, these tests, even with high values, are not enough to guarantee the performance of the seeds in the field, since their potential also depends on the environmental conditions (NASCIMENTO and PEREIRA, 2007).

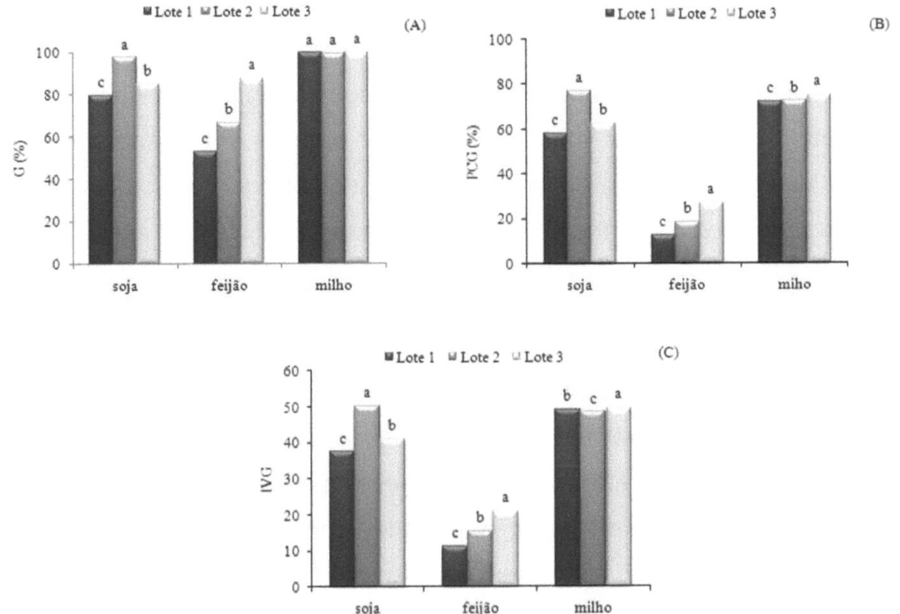

Figure 1: (A) Germination percentage (G), (B) first germination count (PCG) and (C) germination speed index (IVG) of three batches of soybean (*Glycine max* L.), bean (*Phaseolus vulgaris* L.) and maize (Zea *mays* L.) seeds. Averages followed by the same letter do not differ by Tukey's test at 5% probability.

The electrical conductivity (EC) test, which assesses the integrity of cell membranes, showed differences between the lots in terms of vigor for the three crops after three hours of soaking (Figure 2A). The results found for soybean seeds show that the EC reading for batch one (2.1 μS cm^{-1} g^{-1}) was higher when compared to batches two (1.6 μS cm^{-1} g^{-1}) and three (1.82 μS cm^{-1} g^{-1}). In this way, batch two showed lower electrical conductivity values than batches one and three, characterizing the lower release of exudates by the seeds.

In relation to the bean and maize seeds, during the same soaking period, batch three of both species showed greater vigor (Figure 2A), as it showed lower leachate loss (1.68 and 1.87 µS cm^{-1} g^{-1} , respectively) and, consequently, greater integrity of its cell membranes, which is in line with the results of the G, PCG and IVG tests (Figure 1A, 1B and 1C, respectively). In this way, the EC results after three hours of soaking allowed the batches to be separated into different levels of vigor, highlighting the importance of vigor tests in the sense that they show the start of the deterioration process (SANTOS et al., 2005).

After 24 hours of soaking, the results corroborate those obtained after three hours of soaking for the three crops (Figure 2B). It was therefore found that, regardless of the soaking period, the information provided by the test for the three crops was directly related to the tests used to assess the initial quality of the seed lots (Figure 1). Information along these lines was also found by Alves and Sà (2009) with rocket seeds, where they concluded that the electrical conductivity test proved to be efficient for assessing the physiological potential of these seeds. Similarly, the efficiency of the EC test for separating lots was also observed in seeds of canola (*Brassica napus* L.) (AVILA et al., 2005), sunflower (*Helianthu annuus* L.) (BRAZ and ROSSETTO, 2009), turnip rape (*Raphanus sativus* L.) (NERY et al., 2009) and soybean (*Glycine max* L.) (CARVALHO et al., 2009).

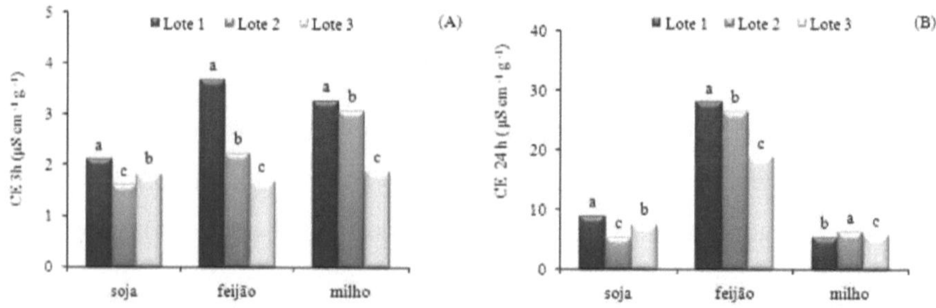

Figure 2 (A,B) Electrical conductivity (EC) at three and 24 hours of incubation of three batches of soybean (*Glycine Max* L.), bean (*Phaseolus vulgaris* L.) and corn (Zea *Mays* L.) seeds. Averages followed by the same letter do not differ by Tukey's test at 5% probability.

According to the analyses relating to the length of the aerial part (CPA) and roots (CR) of the soybean seedlings from the germination test (Figures 3A and 3B), in the same way as G, PCG, IVG and CE (Figures 1 and 2, respectively), batch two was significantly superior to the others, while batch one performed worse. There were no significant differences between the three lots in terms of the CPA and CR of the bean and maize seedlings (Figures 3A and 3B). However, in work carried out with soybean seeds, it was found that CR was efficient for separating lots at different levels of vigor (VANZOLINI *et al.*, 2007).

The MSPA evaluation of the soybean seedlings showed no difference in separating the lots into different levels of vigor (Figure 3C), unlike the MSR (Figure 3D) which distinguished the soybean seed lots into three levels of vigor, as well as the CPA and CR results, with lot two standing out as the most vigorous (Figure 3A and 3B, respectively). This corroborates the results obtained by Nascimento and Pereira (2007), which indicate the efficiency of evaluating primary root length to separate lettuce lots into different levels of vigor.

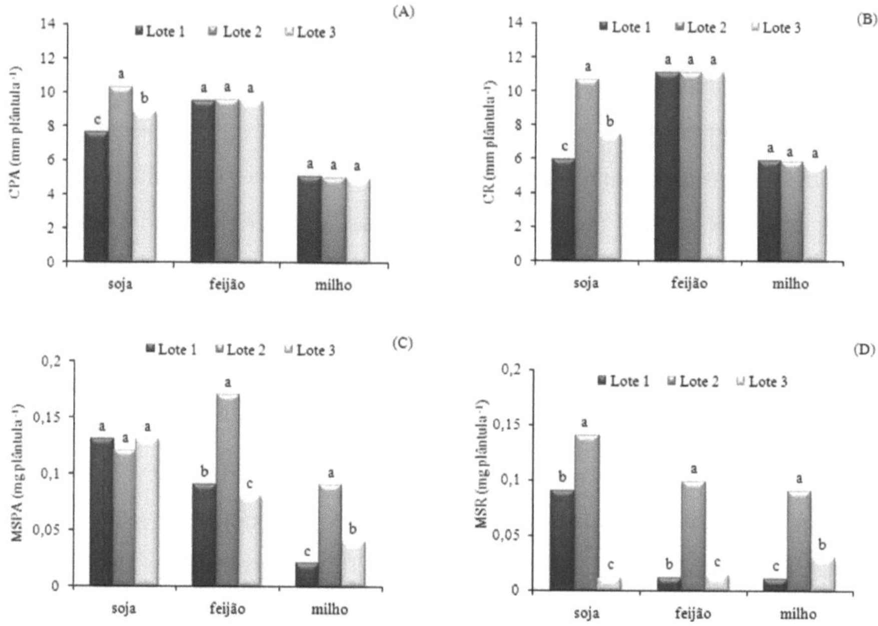

Figura 3. (A, B) Length and dry mass of the aerial part (CPA; MSPA) and (C, D) of the roots (CR; MSR) of seedlings from three batches of soybean (*Glycine max* L.), bean (*Phaseolus vulgaris* L.) and maize (*Zea mays* L.) seeds from the germination test. Averages followed by the same letter do not differ by Tukey's test at 5% probability.

The MSPA and MSR variables of bean and maize seedlings differentiated the three seed lots into different levels of vigor (Figure 3C and 3D, respectively), showing that lot two had the highest biomass accumulation in both crops.

For seedling emergence in the greenhouse, according to the results found for the batches of soybean and bean seeds, it was found that E and IVE (Figure 4A and 4B, respectively) corroborate the results found in the viability and vigor tests described above (Figures 1 and 2). However, for maize seeds, no significant difference was observed between batches in terms of emergence (Figure 4A), which was only observed for IVE, indicating the superiority of batch one over the other batches (Figure 4B).

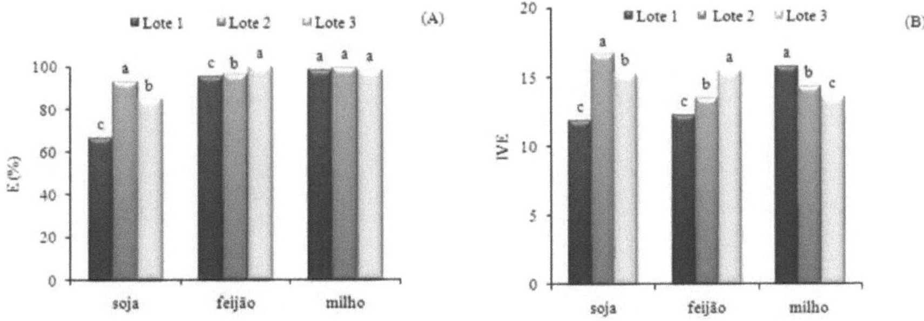

Figura 4. (A) Seedling emergence and (B) emergence speed index (ESI) of three batches of soybean (*Glycine max* L.), bean (*Phaseolus vulgaris* L.) and corn (*Zea mays* L.) seeds. Averages followed by the same letter do not differ by Tukey's test at 5% probability.

This result is not in line with the other variables analysed, which characterized batch three as having the best vigour, however, this response may have been influenced by the environmental adversities to which the IVE test is exposed, since in this case there is no control over external factors.

In the same way as E and IVE, the length of the aerial part (CPA) and roots (CR), as well as the dry mass of the aerial part (MSPA) and roots (MSR) of the three batches of soybean and bean seeds grown in the greenhouse, were efficient in distinguishing the batches, confirming batch two as having high vigor for soybeans and batch three for beans (Figures 5). However, these same variables did not make it possible to differentiate the maize batches (Figures 5), which can be explained by the high germination percentage for this crop, showing similar responses for most of the observed variables, which made it difficult to separate these batches.

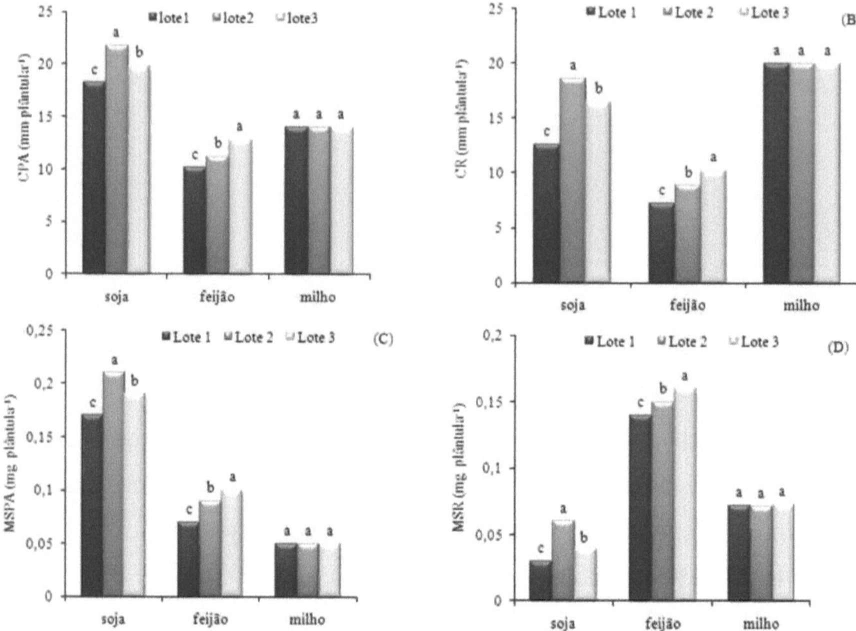

Figura 5. (A) Aerial part length (CPA) and (B) root length (CR); (C) aerial part dry mass (MSPA) and (D) root dry mass (MSR) of seedlings from the three lots of soybean (*Glycine max* L.), bean (*Phaseolus vulgaris* L.) and maize (*Zea mays* L.) seeds from the emergence test in the greenhouse. Averages followed by the same letter do not differ by Tukey's test at 5% probability.

Given these results, it was possible to relate the responses found for the seed lots of the three crops to their respiratory activity measured in the Pettenkofer apparatus. For soybean seeds, this test indicated high respiratory activity for batch two, characterizing it as having the highest vigour (Figure 6). This response suggests that the activity and integrity of the mitochondria of viable embryos increases from the start of soaking, making the production of ATP, which is the energy required for plant development, more efficient (CASTRO and HILHORST, 2004).

36

These results are in line with the classification of the three batches of soybeans observed in the standard viability and vigor tests (Figures 1, 2 and 3). The same was found for batches of sunflower seeds cv. MG2 (DODE et al., 2012), pigeonpea (AUMONDE et al., 2012) and soybean cv. 8000 (MENDES et al., 2009), which corroborates the results of this study, which showed this method to be an important tool in separating seed lots in terms of vigor for various crops.

For bean seeds, in the same way as for soybean seeds, the respiration results were reinforced by the results obtained in the other standard viability and vigor tests (Figures 1, 2 and 3), with emphasis on the EC test, which in both soaking periods showed a tendency for leachate loss in the bean seed batches of lower vigor, showing a greater speed in the process of membrane deterioration.

However, it is important to note that, in general, the standard seed physiological quality tests did not make it possible to differentiate between batches of maize seeds, as they were all highly viable. However, when the seed lots were evaluated by measuring respiration activity, it was possible to detect slight differences in relation to vigor between the lots, which demonstrated the high efficiency of this test.

Studies show that the changes responsible for the drop in vigor reduce the respiration rate and the activity of enzymes, while others suggest a reduction in the quantity of enzymes and the normality of their formation in the mitochondria (MENDES et al., 2009). Due to the fact that this organelle in dry seeds and at the beginning of the soaking process does not have an organized membrane system, structural recovery begins to occur as hydration proceeds, becoming more efficient in oxidative phosphorylation (CASTRO and HILHORST, 2004).

Therefore, the maintenance of vigor can be seen as a consequence of the period of time needed for the mitochondria to become more efficient, to start performing respiratory functions and for the membrane system to become more organized (MARCOS FILHO et al., 2006).

As observed in this study, the greatest respiratory activity was seen in the batches of seeds that were classified as the most vigorous, through the standard physiological quality tests, given that the greater release of CO_2 characterized the integrity of the cell membranes, including the mitochondrial membranes, with the beginning of the respiratory process being recorded, which is the first metabolic activity that is rapidly activated immediately after the seeds are soaked, initiating the germination process (HOFS et al., 2004).

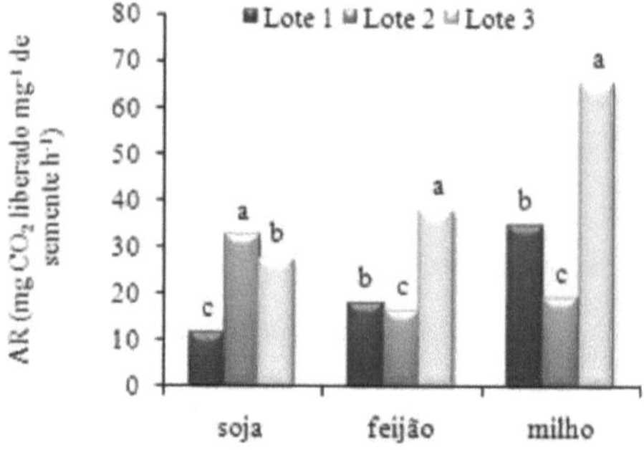

Figura 6. Respiratory activity (RA) of three batches of soybean *(Glycine max* L.), bean (*Phaseolus vulgaris* L.) and maize (*Zea mays* L.) seeds measured in the Pettenkofer Apparatus. Averages followed by the same letter do not differ by Tukey's test at 5% probability.

This study concludes that respiratory activity is an efficient technique for separating batches of soybean, bean and corn seeds in terms of vigor.

REFERENCES

ABRANTES,F.L.; KULCZYNSKI,S.M.; SORATTO,R.P.; BARBOSA,M.M.M. Nitrogen in top dressing and physiological and health quality of *millet* seeds (*Panicum miliaceum* L.). **Revista Brasileira de Sementes** ,v.32, p.106-115, 2010.

ALVES, C.Z.; SA, M.E. Electrical conductivity test to evaluate the vigor of arugula seeds. **Revista Brasileira de Sementes**, v.31, p.203-215, 2009.

AUMONDE, T.Z.; MARINI, P.; MORAES, D.M.;MAIA, M.S.; PEDÓ, T.;TILLMANN, M.A.A.;VILLELA, F.A. Classification of the vigor of kidney bean seeds by respiratory activity. **Revista Interciência**, v. 37, p. 55-58, 2012.

AVILA, M.R.; BRACCINI, A.I.; SCAPIM, C.A.;MANDARINO, J.M.G.; ALBRECHT, L.P.; VIDIGAL FILHO, O.P. Yield components, isoflavone, protein and oil content and quality of soybean seeds. **Revista Brasileira de Sementes**, v.29, p. 111-127, 2007.

AVILA, M.R.; BRACCINI, A.L.; SCAPIM, C.A.; MARTORELLI, D.T.; ALBRECHT, L.P. Laboratory tests on canola seeds and correlation with seedling emergence in the field. **Revista Brasileira de Sementes**, v. 27, p. 62-70, 2005.

BARBIERI, A.P.P.; MENEZES, N.L.; CONCEIÇÂO, G.M.; TUNES, L.M. Potassium leaching test for the evaluation of rice seed vigor. **Revista Brasileira de Sementes**, v. 34: 117-124, 2012.

BRASIL (2009) **Rules for Seed Analysis**. SNAD/CLAV. Ministry of Agriculture and Agrarian Reform. Brasilia, Brazil. 398p.

BRAZ, M.R.S.; ROSSETTO, C.A.V. Correlation between tests for evaluating sunflower seed quality and seedling emergence in the field. **Ciência Rural Magazine**, v. 39, p.2004-2009, 2009.

CARVALHO, L.F.; SEDIYAMA, C.S.; REIS, M.S.; DIAS, D.C.F.S.; MOREIRA, M.A. Influence of soybean seed soaking temperature on the electrical conductivity test to evaluate physiological quality. **Revista Brasileira de Sementes**, v.41, p.9-17, 2009.

CASTRO, R.D.; HILHORST, H.W.M. Soaking and reactivation of metabolism. In: Ferreira AG, Borghethi F(Orgs). **Germination from basic to applied**. Porto Alegre: Artmed, 149-162, 2004.

CONAB (NATIONAL SUPPLY COMPANY) 2013. Monitoring the Brazilian grain harvest. Fifth survey/Brasilia, Brazil. 28p.

DODE, J.S.; MENEGHELLO, G.E.; MORAES, D.M.; PESKE, S.T. Respiration test to evaluate the physiological quality of sunflower seeds. **Revista Brasileira de Sementes**, v.34, p. 686-691, 2012.

DUTRA, A.A.; MEDEIROS FILHO, S.; DINIZ, F.O. Germination of albizia seeds (*Albizia lebbeck* (L.) Benth.) as a function of light and temperature regime. **Revista Caatinga**, v. 21, p.75-81, 2008.

DUTRA, A.S.; VIEIRA, R.D. Electrical conductivity test to evaluate the vigor of zucchini seeds. **Revista Brasileira de Sementes**, v. 28, p.117-122, 2006.

HOFFS, A.; SCHUCH, L.O.B.; PESKE, S.T.; BARROS, A.C.S.A. Effect of seed physiological quality and sowing density on grain yield and industrial quality in rice. **Revista Brasileira de Sementes**, v. 26, p. 55-62, 2004.

KRZYZANOWSKI, F.C.; FRANÇA-NETO, J.B.; HENNING, A.A. Report on vigor tests available for major crops. **Informativo *ABRATES***, v.1, p. 15-50, 1991.

MACHADO, A.; CONCEIÇÂO, A.R. (2007). **Statistical program winstat: statistical analysis system for windows.** Pelotas, Brazil.

MAGUIRE, J. Speed of germination-Aid in selection and evaluation for seedling emergence and vigor. *Crop Science, v.* 2, p.176, 1962.

MARCOS FILHO, J.; KIKUTI, A.L.P.; LIMA, L.B. Methods for evaluating the vigor of soybean seeds, including computerized image analysis.

Revista Brasileira de Sementes, v. 31, p.102-112, 2009.

MARCOS FILHO, J.; KIKUTI, A.L.P. Radish seed vigor and plant performance in the field. **Revista Brasileira de Sementes**, v. 28, p.44-51, 2006.

MARCOS FILHO, J.; BENNETT, M.A.; MCDONALD, M.B.; EVANS, A.;GRASSBAUGH, E.M. Assessment of melon seed vigour by an automated computer imaging system compared to traditional procedures. **Seed science and technology**, v.34, p.485-497, 2006.

MENDES, C.R.; MORAES, D.M.; LIMA, M.G.S.; LOPES, N.F. Respiratory activity for the differentiation of vigor on soybean seed lots. **Revista Brasileira de Sementes**, v.31, p. 171-176, 2009.

MORAES, D.M.; BANDEIRA, J.M., MARINI, P., LIMA, M.G.S.; MENDES, C.R. (2012). **Laboratory Practices in Plant Physiology**. Pelotas, Brazil, v. 3, 52 p.

NASCIMENTO, W.M.; PEREIRA, R.S. Tests to evaluate the physiological potential of lettuce seeds and its relationship with germination under adverse temperatures. **Revista Brasileira de Sementes**, v. 29, p.175-179, 2007.

NERY, M.C.; CARVALHO, M.L.M.; GUIMARAES, R.M. Vigor tests for evaluating the quality of turnip seeds. **Informativo *Abrates***, v. 19, p.9-20, 2009.

SANTOS, C.M.R.; MENEZES, N.L.; VILLELA, F.A. Physiological and biochemical changes in bean seeds during storage. **Revista Brasileira de sementes**, v.27, p.104-114, 2005.

VANZOLINI, S.; ARAKI, C.A.S.; SILVA, A.C.T.M.; NAKAGAWA, J. Seedling length test in the evaluation of the physiological quality of soybean seeds. **Revista Brasileira de sementes**, v.29, p.90-96, 2007.

yes
I want morebooks!

Buy your books fast and straightforward online - at one of world's fastest growing online book stores! Environmentally sound due to Print-on-Demand technologies.

Buy your books online at
www.morebooks.shop

Kaufen Sie Ihre Bücher schnell und unkompliziert online – auf einer der am schnellsten wachsenden Buchhandelsplattformen weltweit! Dank Print-On-Demand umwelt- und ressourcenschonend produziert.

Bücher schneller online kaufen
www.morebooks.shop

info@omniscriptum.com
www.omniscriptum.com